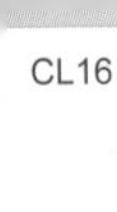

ATOMIC

Nature's Armour AND DEFENCES

PAUL MASON

www.raintreepublishers.co.uk
Visit our website to find out more information about **Raintree** books.

To order:
Phone 44 (0) 1865 888112
Send a fax to 44 (0) 1865 314091
Visit the Raintree bookshop at **www.raintreepublishers.co.uk** to browse our catalogue and order online.

First published in Great Britain by Raintree, Halley Court, Jordan Hill, Oxford OX2 8EJ, part of Harcourt Education. Raintree is a registered trademark of Harcourt Education Ltd.

Editorial: Melanie Waldron and Harriet Milles
Design: Steve Mead, Victoria Bevan, and Bigtop
Illustrations: Jeff Edwards and Darren Lingard
Picture Research: Mica Brancic
Production: Julie Carter

Originated by Chroma Graphics
Printed and bound in China by Leo Paper Group

ISBN 978 1 4062 0675 3 (hardback)
12 11 10 09 08
10 9 8 7 6 5 4 3 2 1

ISBN 978 1 4062 0696 8 (paperback)
12 11 10 09 08
10 9 8 7 6 5 4 3 2 1

British Library Cataloguing in Publication Data
Mason, Paul, 1967-
Nature's armour and defences. - (Atomic)
1. Animal defences - Juvenile literature
591.4'7
A full catalogue record for this book is available from the British Library.

Acknowledgements
The publishers would like to thank the following for permission to reproduce photographs:
Anders Waren p. **10**; Ardea pp. **6** top (Jean Michel Labat), **14** (Bob Gibbons), **27** (Eric Dragesco); Corbis pp. **6** (bottom) (Sunny S. Unal), **21** (Gavriel Jecan), **22** (Michael & Patricia Fogden); Getty Images/PhotoDisc p. **23**; imagequestmarine.com/Masa Ushioda p. **26**; Nature Picture Library pp. **13** bottom (Mark Payne-Gill), **25** (Tom Vezo); OSF pp. **9** (Brian Kenney), **13** top (Animals Animals/Earth Scenes), **17** bottom (David Thompson), **17** top (Pacific Stock/Robinson Ed), **18** (Tim Jackson).

Cover photograph of a cactus reproduced with permission of Corbis. Image of a hermit crab reproduced with permission of NHPA / Martin Harvey.

Every effort has been made to contact copyright holders of any material reproduced in this book. Any omissions will be rectified in subsequent printings if notice is given to the publishers.

The publishers would like to thank Nancy Harris, Dee Reid, and Diana Bentley for their assistance with the preparation of this book.

Contents

Some words are printed in bold, **like this**. You can find out what they mean in the glossary. You can also look in the box at the bottom of the page where the word first appears.

Dangerous World

There is danger everywhere for small animals. Wherever they go, bigger animals are hoping to turn them into breakfast (or lunch, or dinner)! Luckily, animals – and even some plants – have some great ways of protecting themselves.

NORTH AMERICA

ATLANTIC OCEAN

PACIFIC OCEAN

SOUTH AMERICA

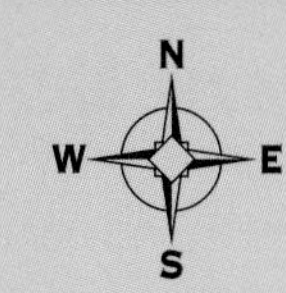

1 Porcupine

Location: *North and South America, Africa, and Asia*

Porcupines use their prickly tails to hit their attackers.

pages 18 and 19

6 Armadillo

Location: *South and North America*

Armadillos have special tricks for making sure attackers cannot get through their armour.

pages 12 and 13

② Poisonous toads and frogs

Location: *Equatorial regions*

If an animal tries to attack a harmless-looking toad or frog, it might get a nasty surprise!

pages 22 and 23

EUROPE

ASIA

MEDITERRANEAN SEA

AFRICA

PACIFIC OCEAN

Equator

INDIAN OCEAN

AUSTRALIA

③ Iron-plated snail

Location: *Indian Ocean*

This tiny snail uses a layer of metal to stay safe from animals that want to eat it!

pages 10 and 11

⑤ Skunk

Location: *North and South America*

Some people say that the skunk has the best defence of all!

pages 24 and 25

④ Hermit crab

Location: *Sea shores around the world*

The cheeky hermit crab uses someone else's armour, instead of growing its own!

pages 16 and 17

This turtle has a tough shell of armour.

Lionfish have poisonous **spines** to protect them from **predators**.

Attackers Everywhere!

For most animals and plants, the danger of being attacked is always there. Somewhere, another creature is hoping to eat them!

Why do animals use armour?

Armour is a strong cover that acts as a **barrier** around an animal's body. It is difficult for enemies to get through the protective barrier to attack the animal inside.

Do animals use other defences?

Armour can be heavy and slow animals down. Some animals have to be fast moving, so they cannot carry around heavy armour. These animals **defend** themselves in other ways.

barrier	**layer of defence**
defend	**protect against attack**
predator	**animal that hunts other animals for food**
spine	**stiff spike that points outwards**

Defence Systems

Animals use armour and defences to make things as tough as they can for predators.

Armoured coats

Tough armour such as a turtle's shell is difficult for predators to bite through. Other animals are armoured with spikes, similar to a porcupine's **spines**.

Other defences

A few animals have horrible-tasting skin. Other creatures do not wait to be bitten; they spray predators with something unpleasant!

Amazing Fact!

Some trees defend themselves with sticky gum. The rubber in rubber trees, for example, glues together the jaws of any animal that bites it!

They may look easy to catch, but these poison arrow frogs could taste terrible!

The iron-plated snail's plates are slightly **magnetic**, because they are so full of iron!

magnetic	**able to pull iron or steel towards it**

The Iron-Plated Snail

Some creatures use armour just like knights did many years ago. This solid barrier stops attackers from hurting them.

Iron armour

Surely animals do not actually use iron suits like a knight's? The iron-plated snail does! This amazing creature lives in deep water. Its shell is made of pieces that lock together, like a jigsaw.

The iron-plated snail uses its armour to stop attacks by poisonous snails. The iron plates prevent the poison from getting through.

Amazing Fact!

Australia is home to a tree with such hard bark that it is called the "ironbark tree". Imagine trying to chew into that!

Armoured Armadillos

Some animals can grow their own armour because it is made of bone. Biting through bone is very difficult. Most predators find it impossible!

Armadillo armour

Armadillos have armour on their back and sides. Most armadillos also have armour on their heads and tails. Only their undersides have no armour. This is where predators can bite them.

Defending the underside

How do armadillos **defend** their soft underside? Many grip the ground tightly, digging their armour into the earth. This makes it harder for predators to tip them over.

Amazing Fact!

Armadillos often leap up in the air if they are startled.

The three-banded armadillo can curl up into an armoured ball.

This armadillo can grip the ground like a limpet when threatened. Turn the page to find out more about limpets.

Not many predators can get through the limpets' gripping defence.

Amazing Fact!

After feeding, limpets always return to the exact same place on their rock. No one knows how they find their way back!

GRIPPY LIMPETS

The armadillo isn't the only animal that grips on tight when attacked. Some sea creatures grip on even harder than an armadillo!

Armoured shells

Limpets live on rocks in cool seawater. They crawl about during the day, searching for food. If **predators** approach, the limpet hides under an armoured shell that it carries on its back.

If a predator pulled the limpet's shell off, it could easily be eaten. Fortunately for the limpet, it is famous for its gripping ability! "Cling on like a limpet" is a saying that means "Hold on very tight".

When threatened, limpets hide under their shell.

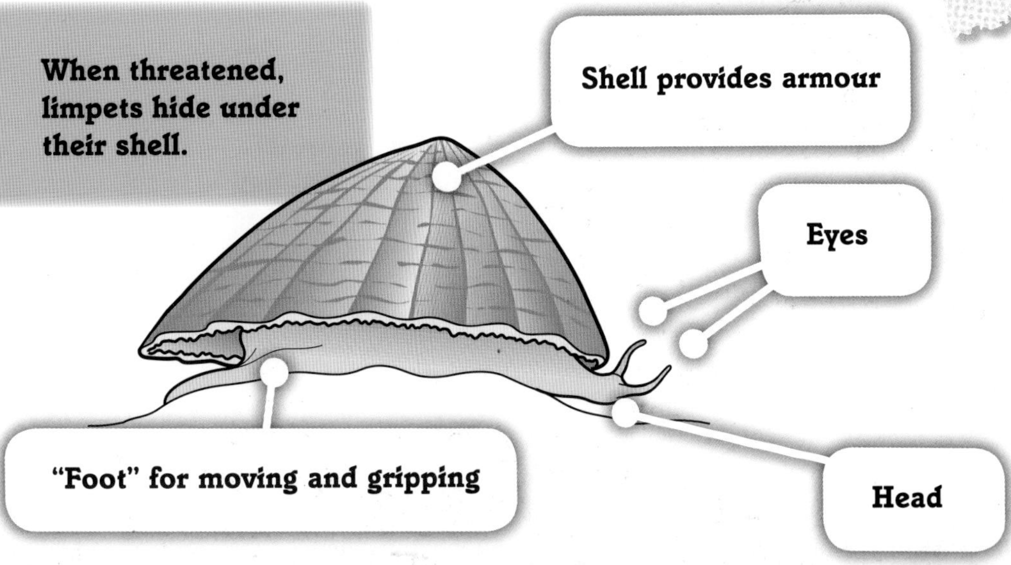

Shell provides armour

Eyes

"Foot" for moving and gripping

Head

Hideaway Hermit Crabs

Some animals do not bother with making their own armour. Instead, they use armour that someone else has left lying around!

Abandoned armour

Hermit crabs use the **abandoned** shells of sea snails as armour. The crab's soft body fits into the twists of the shell. The hermit crab carries the shell around and hides inside when **predators** threaten.

Moving house

Sometimes, a hermit crab grows too big for its shell. It has to move house, to a bigger shell.

Amazing Fact!

Most hermit crabs have a right claw much bigger than their left. This big claw is a useful weapon!

abandon leave behind

The hermit crab borrows the shells of dead sea snails.

Hermit crabs are in great danger while they move to a new shell. For a little while, they don't have any armour!

Lioness beware! These porcupines would make a very painful mouthful!

Prickly Porcupines

Having a spiky coat is a good defence against predators! A few animals grow sharp spines. Their spines stick into attackers and break off.

Porcupine spines

A porcupine's spines are actually hairs! They are stiff and hollow. This makes them light enough for the porcupine to move around quickly.

When predators approach, the porcupine turns away from them. Then it lashes out with its spiny tail. If the spines break off, they work their way into the predator's skin. This is very uncomfortable!

Amazing Fact!

A large porcupine has a lot of spines. There may be as many as 30,000! But porcupines do not have any spines on their stomachs.

Spiky Cactus

It's not only animals that use spines to protect themselves from predators. Some plants grow spines as a form of defence, too!

Desert water store

Cactus plants grow in **deserts** and dry areas. They have thick, waxy skin and a spongy inside. The cactus is very good at storing water.

Water hunters

Thirsty animals would like to get at the cactus's water. It needs to **defend** itself against attack.

The cactus has spines instead of leaves. The spines are sharp and thin. If an animal tries to eat the cactus, the spines break off and stick into its skin. This is itchy and painful!

desert dry area where very little rain falls

Cactus spines have other uses, apart from defence. They give the plant a little shady protection from the hot desert sun.

Poisonous frogs often have vivid markings. This warns **predators** that they might taste nasty!

captivity **kept in a cage or tank**

Terrible Toads and Poison Frogs

Some animals have an invisible type of armour. They use a terrible taste or stinky smell to protect themselves.

Terrible toads

Some toads have special skin defences. When they are threatened, their skin releases a chemical. The chemical tastes foul. So an animal might think it is about to have a tasty snack, but it ends up spitting the toad out in disgust!

Poison frogs

Some frogs do not simply taste bad, they are also poisonous! If an animal tries to bite one of these frogs, it could end up dead.

Amazing Fact!

Most poison frogs lose their poison if they are kept in captivity. Scientists think this is because they don't eat the same foods as they would in the wild.

Stinky Skunks

Skunks do not use armour that you can see. Instead they use a stinky smell to keep themselves safe!

Stripy stinkers

The most famous stinker is the striped skunk. The big white stripes on the skunk's fur coat are a warning to **predators** to stay away – or else!

Pooh!

If a predator gets too close, the skunk stamps its feet as a warning. If this does not work, the skunk turns around, lifts its tail, and "fires" a foul-smelling stream at the predator's face.

Amazing Fact!

The spotted skunk can spray predators while doing a handstand! This gives it a better aim.

This stinker is ready to fire a defensive pong if threatened.

This squid has squirted ink into the water. This should help it to escape.

Amazing Fact!

Giant squid can grow to over 13 metres (43 feet) long!

Squirty Squid

Some animals have more than one defence. If their first defence does not work, they have a back-up plan!

Colourful camouflage

Squid protect themselves from **predators** by using **camouflage**. They change the colour of their body to blend in with the background. This makes it more difficult for predators to see them.

Disappearing into the darkness

Sometimes a predator does spot a squid. Then the squid has a second defence. It squirts a large cloud of ink into the water. The water goes dark, preventing the predator from seeing the squid, who then makes its escape!

camouflage way of hiding or disguising yourself by looking like your surroundings

The Best Defences

Animal defences include armour, spikes, and poisonous skin. But which animal has the best defences of all?

The honey badger

The honey badger has **multiple** defences:

- First of all, it can make a terrible **odour**, like a skunk. If that does not put off the attacker...
- The badger's skin is tough and rubbery. It is difficult for **predators** to get their teeth into. Even if they manage it...
- The badger's skin is very loose. The badger can wriggle around inside its skin, before reaching round to bite its attacker.

Amazing Fact!

The honey badger is in the *Guinness Book of Records* as the "world's most fearless animal". These badgers have been known to chase off lions and leopards.

multiple many
odour smell

Honey badgers are very bad-tempered.

Glossary

abandon leave behind

barrier layer of defence. Barriers are often solid, like a wall or fence.

camouflage way of hiding or disguising yourself by looking like your surroundings

captivity kept in a cage or tank. Pets and zoo animals are kept in captivity.

defend protect against attack. Animals have lots of different ways of defending themselves including clouds of ink and strong smells.

desert dry area where very little rain falls. Desert plants and animals have to survive on small amounts of water.

magnetic able to pull iron or steel towards it. A fridge magnet is an example of something magnetic that we use in everyday life.

multiple many; more than one

odour smell

predator animal that hunts other animals for food

spine stiff spike that points outwards. It is usually at the end of a long, thin rod or tube.

Want to Know More?

Books

- *Amazing Nature: Powerful Predators*, Tim Knight (Heinemann Library, 2005)
- *Amazing Nature: Super Survivors*, Tim Knight (Heinemann Library, 2005)
- *Animal Camouflage and Defence*, Kate Petty (Heinemann Library, 2004)
- *Predators and Prey*, Michael Chinery (Zero to Ten, 2000)

Websites

- www.honeybadger.com
  Find out about the world's most fearless animal and its amazing defences.
- www.nhm.ac.uk
  The website of the Natural History Museum, London.
- www.amnh.org
  The website of the American Museum of Natural History.

If you liked this Atomic book, why don't you try these...?

Index

Notes for adults

Use the following questions to guide children towards identifying features of report text:

Can you find an example of a general opening classification on page 7?
Can you give an example of a generic participant on page 8?
Can you find examples of non-chronological language on page 11?
Can you find examples of the details of a limpet on page 15?
Can you give examples of present tense language on page 23?